WHAT IF YOU HAD AN
Animal Tongue!?

by **Sandra Markle**

Illustrated by
Howard McWilliam

Scholastic Inc.

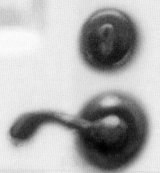

For Katie Kibler and the children
of Avery Elementary School
in Hilliard, Ohio.

A special thank-you to Skip
Jeffery for his loving support
during the creative process.

The author would like to thank the following people for sharing their enthusiasm and expertise: Dr. Stephen M. Deban, University of South Florida, Tampa, Florida; Dr. Bryan Fry, University of Queensland, St. Lucia, Queensland, Australia; Dr. Abdul Haleem, Aligarh Muslim University, Uttar Pradesh, India; Dr. Nathan Muchhala, University of Missouri, St. Louis, Missouri; Dr. Doug Smith, project leader for the Yellowstone Gray Wolf Recovery Project, Yellowstone National Park, Wyoming; Erin Stahler, biological technician and program manager for the Yellowstone Wolf Project, Yellowstone National Park, Wyoming; Dr. Martin Whiting, Macquarie University, Sydney, Australia.

Photos ©: cover bottom right: Murray Cooper/Minden Pictures; 4: Andreygudkov/Getty Images; 4 inset: Mikel Bilbao/VW PICS/Universal Images Group/Getty Images; 6: Murray Cooper/Minden Pictures; 6 inset: Murray Cooper/Minden Pictures; 8: Juniors Bildarchiv GmbH/Alamy Stock Photo; 8 inset: RudyBalasko/Getty Images; 10: Rauschenbach, F./picture alliance/Arco Images G/Newscom; 10 inset: Dr. Stephen M. Deban; 12 inset: Andra Boda/EyeEm/Getty Images; 14 inset: Tierfotoagentur/Alamy Stock Photo; 16: Joseph T. Collins/Science Source; 18: Gary Carter/Getty Images; 18 inset: ZUMA Press Inc/Alamy Stock Photo; 20: Stefan Huwiler/Getty Images; 20 inset: Michael Cummings/Getty Images.

All other photos © Shutterstock.com.

Library of Congress Cataloging-in-Publication Data available

ISBN 978-1-338-59668-7

10 9 8 7 6 5 4 3 2   20 21 22 23 24

Printed in China   38
First edition, September 2020

Book design by Kay Petronio

What if one day when you woke up, you felt a little bit strange? And you discovered that the tongue in your mouth was now VERY different! What if, overnight, a wild animal's tongue had taken its place?

KOMODO DRAGON

A Komodo dragon sticks out its long, yellow tongue to snag air samples. Pressing its tongue to a special sensor on the roof of its mouth lets it smell/taste what it collected. It's how the dragon locates the meat it eats, such as deer, pig, or water buffalo, which it senses as far as five miles away.

FACT

A Komodo dragon's tongue tip is forked to sense if the smell/taste is stronger to the left or the right.

If you had a Komodo dragon's tongue, you'd never miss out on freshly baked cookies.

TUBE-LIPPED NECTAR BAT

A tube-lipped nectar bat's tongue is one and a half times longer than its body—so long that it's attached inside its rib cage instead of its mouth. Stretched out, its tongue is long enough to feed on the sweet nectar in the bottom of tubular flowers that are too long for short-tongued bats to reach.

FACT

A tube-lipped nectar bat's tongue has a prickly covering of hairlike bristles—perfect for trapping nectar drops and holding them all the way to its mouth.

If you had a tube-lipped nectar bat's tongue, you'd never have to pause the movie.

TIGER

A tiger's tongue is sandpapery-rough with backward-facing spines that flex and turn with each lick. These comb fuzzy snarls and knots from its fur. Spending as much as a quarter of its waking time licking keeps the tiger one well-groomed big cat.

FACT

A tiger's tongue spines are cone-shaped and curved backward toward its throat—just right for picking up a drink one lick at a time.

If you had a tiger's tongue, you'd be a world-famous hair stylist.

WEB-TOED SALAMANDER

A web-toed salamander's tongue shoots out of its mouth so fast it can do ten tongue punches in the time it takes a human to blink once. Scientists believe muscles wrapped around the salamander's tongue are what power this action, like pulling a rubber band so far it snaps. POW!

FACT

When fully extended, a web-toed salamander's tongue is about half its body length (minus the tail).

If you had a web-toed salamander's tongue, you'd be the one to bust open the party's piñata.

11

OKAPI

An okapi's tongue is very handy. At mealtimes, its tongue wraps around and pulls in mouthfuls of leaves, fruit, or twigs. At bath time, its tongue licks its body clean—even its eyes and big ears.

FACT

An okapi's tongue is purplish black. Scientists believe this color protects the tongue from getting sunburned.

If you had an okapi's tongue, you could safely pick apples from the tree's highest branches.

CHAMELEON

A chameleon's tongue has a suction cup tip coated with saliva so gooey it's 400 times thicker than human spit. No wonder whatever a chameleon snags with its tongue stays caught!

FACT

A chameleon's saliva is so sticky that its tongue can haul really big prey into its mouth without losing its grip.

If you had a chameleon's tongue, you'd never miss catching a Frisbee.

15

ALLIGATOR SNAPPING TURTLE

An alligator snapping turtle's tongue tip is topped with a red, worm-shaped lump. When the turtle opens its mouth and wiggles its tongue, the wormy part seems alive. When fish can't resist taking a closer peek, the turtle snaps its mouth shut.
Dinnertime!

FACT

An alligator snapping turtle hatchling already has a fake worm on its tongue ready to trick its

If you had an alligator snapping turtle's tongue, you'd be your soccer team's Most Valuable Player.

RED-BELLIED WOODPECKER

A red-bellied woodpecker's tongue can stretch out nearly three times longer than its beak. Stiffened by bones all the way to a barbed tip, it spears its meals, such as wood-boring beetle larvae.

FACT

A red-bellied woodpecker's tongue tip has special muscles that move it left, right, and around, helping it remove speared food from tree-trunk tunnels.

If you had a red-bellied woodpecker's tongue, you'd be the carnival's balloon-popping champ.

19

WOLF

A wolf's tongue keeps it cool without wetting its coat and chilling its skin. Instead of sweating, a wolf pants, forcing air to flow over its sloppy, saliva-wet tongue. As the saliva dries, this acts like air conditioning to cool the wolf off—even when it's running flat out.

FACT

In wolf groups, called packs, tongue licks are a way to show respect to the alphas—the pack leaders.

If you had a wolf's tongue, you'd win the Ironkids Triathlon without breaking a sweat.

BLUE-TONGUED SKINK

When a blue-tongued skink sticks its tongue out, it means *BOO!* It does this when an enemy, such as a brown falcon, comes close. With luck, the colorful tongue startles the hunter long enough for the skink to escape.

FACT

Scientists discovered that a blue-tongued skink's tongue is brighter farther back in its mouth. So if an enemy keeps coming, the skink opens wider to be even more startling—*hopefully!*

If you had a blue-tongued skink's tongue, dentist appointments would always be a surprise.

GAUR

A gaur's tongue, like its farm cow cousin's, has as many as 25,000 taste buds—more than twice as many as a human tongue. The gaur is strictly a plant-eater, and all those taste buds let it quickly sense plants it can safely eat.

FACT

A gaur uses tongue-licks to say *you're family*. So mating pairs lick each other and mothers lick their calves.

If you had a gaur's tongue, you'd be the taste tester for new ice-cream flavors.

A wild animal tongue could be fun for a while. But you don't need your tongue to catch Frisbees, pick apples, or pop balloons. And you don't need your tongue

to style hair or cool off. So, if you could keep a wild animal's tongue for more than a day, which kind would be right for you?

Luckily, you don't have to choose. The tongue in your mouth will always be a human tongue. It's what you need to taste, chew, spit, and swallow. Plus, it's what you need to whistle, sing, and talk.

Whatever you do, keep your tongue in your mouth because that's polite in most of the world. But if you visit Tibet, stick your tongue out whenever you meet someone. There, it's the way to say "hello."

WHAT'S SPECIAL ABOUT YOUR TONGUE?

You may be surprised to learn your tongue print is as unique to you as your fingerprints. Your tongue is also very flexible because it's all muscle—in fact, it's made up of eight muscles working together. They're the only muscles in your body that move without shifting any of your bones.

Your tongue has between 2,000 and 4,000 taste buds. These are buried inside papillae, the little bumps you can see on top of your tongue. Each papilla has six to eight taste buds packed with sensory cells. People used to believe different areas on the tongue were able to detect different tastes: sweet, sour, salty, and bitter. Now scientists know these tastes are detected all over the tongue.

KEEP YOUR TONGUE HEALTHY

Your tongue needs to be in good condition for you to be healthy. So here are some tips for taking care of your tongue.

- Brush it gently with your toothbrush. One of the causes of bad breath is bacteria (microscopic living things) collecting on your tongue.

- Drink plenty of water every day—at least a half gallon. Your body needs water to produce saliva, the liquid in your mouth that naturally washes away harmful bacteria that can cause bad breath and tooth decay.

- Eat a healthy diet, including foods such as onions, ginger, and coconut, which are noted for helping to fight harmful bacteria.

- Wear a mouth guard that's been custom made for you to protect your tongue as well as your teeth when you play sports, such as football and baseball.

OTHER BOOKS IN THE SERIES